DOW'S
CHEMICAL EXPOSURE INDEX
GUIDE

FIRST EDITION

a *AIChE technical manual* published by
the American Institute of Chemical Engineers
345 East 47th Street, New York, NY 10017

ACKNOWLEDGMENT

The American Institute of Chemical Engineers wishes to acknowledge the cooperation of The Dow Chemical Company in releasing the *Chemical Exposure Index (CEI) Guide* for publication. Special thanks to R. Dauwe, J. T. Marshall, J. C. Thayer, C. A. Schiappa, Jr., A. G. Mundt, J. F. Murphy, N. E. Scheffler, S. M. Hartnagle, N. H. Humphrey, B. H. Scortichini and T. O. Gibson of The Dow Chemical Company, whose careful evaluation and review of this material provided some important innovations to make the *Guide* more useful to the chemical industry.

AIChE Publications Staff

Managing Editor: Haeja L. Han
Assistant Editor: Arthur Baulch
Art Director: Joseph Roseti

TABLE OF CONTENTS

PREFACE: CHEMICAL EXPOSURE INDEX GUIDE

Background

As a result of various petrochemical incidents occurring in the mid-1980s, the *Chemical Exposure Index Guide* (CEI) was developed and published internally at The Dow Chemical Company in May 1986. Along with the *Fire and Explosion Index Hazard Classification Guide* (F&EI), these two guides have served as a relative ranking analysis to evaluate the hazard potential of installations or changes to installations associated with our facilities.

Corporate Minimum Requirements

The CEI and F&EI guides are a requirement of the Dow Corporate *Minimum Requirements* for Safety, Loss Prevention and Security (June 1989, 4th Edition).

Industry and Governmental Availability

The CEI and F&EI have been made available to all interested parties through the American Institute of Chemical Engineers (AIChE), 345 East 47th Street, New York, NY 10017 (Phone 212-705-7657). In fact, various countries (including the Netherlands and the United States) have referenced the F&EI and CEI guides in their respective governmental regulations.

Improvements Made to This Guide

This guide was prepared incorporating the following improvements:

1. The new CEI methodology utilizes a linearized expression (rather than a step function) for estimating airborne quantity released. The new methodology provides an index that is consequence-based and is independent of the frequency of events. This results in an index that is suited for use as a screening tool for more sophisticated process hazard analyses. These analyses are outlined in the Corporate *Process Risk Management Guidelines for Facilities and Distribution*. A CEI greater than 200 for facilities will require further risk review.

2. Since the CEIs will now function as a screening tool for further analysis, it was necessary to standardize the scenario selection to ensure consistency on a global corporate basis.

3. A comprehensive review process has been included.

This guide is a completely revised document including a new methodology for determining the CEI. Because the new methodology is continuous in many of the variables, the scale for the new CEI is completely different from the earlier version. *CEI values from the earlier version cannot be compared to the CEI values calculated by the procedure described here.*

INTRODUCTION: CHEMICAL EXPOSURE INDEX

The Chemical Exposure Index (CEI) provides a simple method of rating the relative acute health hazard potential to people in neighboring plants or communities from possible chemical release incidents. Absolute measures of risk are very difficult to determine, but the CEI system will provide a method of ranking one hazard relative to another. It is **NOT** intended to define a particular design as safe or unsafe.

All Dow facilities storing or handling acutely toxic materials are expected to calculate the Chemical Exposure Index. This includes new projects as well as existing facilities.

The CEI is used:

- For conducting an initial Process Hazard Analysis (PHA).
- In the Distribution Ranking Index (DRI) calculations.
- By all locations in their review process, which provides the opportunity to make recommendations for eliminating, reducing or mitigating releases.
- In Emergency Response Planning.

Please note that the flammability and explosion hazards are not included in this index. There are other process hazard analysis methods used to identify and measure flammability and explosion hazards. (Refer to AIChE publication, *Fire and Explosion Index Hazard Classification Guide*.)

PROCEDURE FOR
CHEMICAL EXPOSURE INDEX (CEI) CALCULATIONS

1. To develop a Chemical Exposure Index (CEI), the following items are needed:
   a. An accurate plot plan of the plant and the surrounding area.
   b. A simplified process flow sheet showing containment vessels, major piping and chemical inventories.
   c. Physical and chemical properties of the material being investigated, as well as the ERPG/EEPG (pages 5–7) values.
   d. A CEI Guide (Second edition).
   e. A CEI Form (page 20).

   Figure 1 (page 3) is a schematic overview of the CEI calculation. This chart will be helpful as you proceed through this guide.

2. Identify on the process flow sheet any process piping or equipment that could contribute to a significant release of an acutely toxic chemical.

3. Determine the Chemical Exposure Index and the Hazard Distances as explained in the following pages of this guide.

4. Fill out CEI Summary Sheet (page 20).

FIGURE 1

PROCEDURE FOR CALCULATION OF
CHEMICAL EXPOSURE INDEX (CEI)

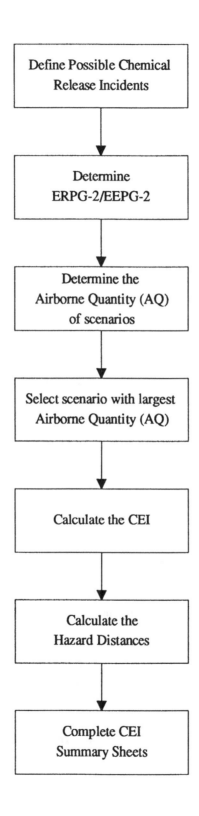

SCENARIOS FOR ESTIMATION OF AIRBORNE RELEASE RATES

The purpose of scenario selection is to determine which process piping or equipment has the greatest potential for the release of significant quantities of acutely toxic chemicals. Since the CEI now serves as a screening tool for further process hazards analysis, it is important that the calculations be done consistently on a global basis. The scenario selection process for determining airborne release rate has, therefore, been standardized to help achieve this goal. The selections listed below were chosen to include the most probable credible events based on historical performance of the chemical industry.

Evaluating several scenarios will aid in determining the largest potential airborne release. Process conditions such as temperature, pressure and physical state should be considered as well as pipe size since they have a significant impact on airborne release rates.

> **Scenario Selection for CEI:** *Select the scenario that gives the largest airborne release.*

1. **PROCESS PIPES**
   Rupture of the largest diameter process pipe as follows:
   - For smaller than 2-inch diameter — full bore rupture
   - For 2- through 4-inch diameter — rupture equal to that of 2-inch diameter pipe
   - For greater than 4-inch diameter — rupture area equal to 20% of pipe cross section area

2. **HOSES**
   Full bore rupture

3. **PRESSURE RELIEF DEVICES RELIEVING DIRECTLY TO THE ATMOSPHERE**
   Calculated total release rate at set pressure. Refer to pressure relief calculation or contact process engineering. All material released is assumed to be airborne.

4. **VESSELS**
   Rupture based on largest diameter process pipe attached to the vessel using pipe criteria above.

5. **TANK OVERFLOWS AND SPILLS**

6. **OTHERS**
   Scenarios can be established based on the plant's or technology's experience, they can be the outcome of a review or derived from hazard analysis studies. They can also be based on the experience of another technology if the event could occur in this unit. Contact Process Engineering for special cases that may include reactivity or mixtures.

The treatment of instantaneous and very short duration continuous releases is simplified for the CEI calculation. Release from all scenarios are assumed to continue for at least a five minute duration. If a release is instantaneous or exceeds the total inventory within this duration, the release rate is calculated by dividing the total inventory by five minutes.

> *After this evaluation, choose the largest airborne release rate for the CEI calculation (page 20).*

EMERGENCY RESPONSE PLANNING GUIDELINES (ERPG) AND DOW EMERGENCY EXPOSURE PLANNING GUIDELINES (EEPG)

The American Industrial Hygiene Association (AIHA) has published Emergency Response Planning Guidelines (ERPG) values which are intended to provide estimates of concentration ranges where one might reasonably anticipate observing adverse effects.

These guidelines are intended to be used as a planning tool for various Dow programs to determine priority concerns, to evaluate the adequacy of containment, to identify downwind areas which might need to take action during a release and to develop community emergency response plans. The need for an ERPG is based on the volatility of a chemical, its toxicity, the releasable quantity and the public's perception of the potential hazard.

The Emergency Exposure Planning Guidelines (EEPGs) are the Dow equivalent to the AIHA published ERPGs. These are provided when AIHA ERPGs do not exist. ERPG/EEPG definitions are as follows:

ERPG-1/EEPG-1 is the maximum airborne concentration below which it is believed that nearly all individuals could be exposed for one hour without experiencing other than mild transient adverse health effects or perceiving a clearly objectionable odor.

ERPG-2/EEPG-2 is the maximum airborne concentration below which it is believed that nearly all individuals could be exposed for up to one hour without experiencing or developing irreversible or other serious health effects or symptoms that could impair their abilities to take protective action.

ERPG-3/EEPG-3 is the maximum airborne concentration below which it is believed that nearly all individuals could be exposed for up to one hour without experiencing or developing life-threatening health effects.

TABLE 1

EMERGENCY PLANNING GUIDELINES: ERPGs/EEPGs

Material	Molecular Weight	Boiling Point °C	ERPG-1 mg/m^3	ERPG-1 PPM	ERPG-2 mg/m^3	ERPG-2 PPM	ERPG-3 mg/m^3	ERPG-3 PPM
Acetone cyanohydrin *	85.11	95			35	10		
Acrolein	56.06	52.5		0.1	1	0.5	7	3
Acrylic acid	72.06	141.4	6	2	147	50	2210	750
Acrylonitrile *	53.06	77.2			43	20		
Allyl chloride	76.53	44.8	9	3	125	40	939	300
Ammonia	17.03	-33.3	17	25	139	200	696	1000
Bromine	159.81	58.7	1	0.2	7	1	33	5
Butadiene	54.09	-4.41	22	10	111	50	11060	5000
n-butyl acrylate	128.17	147.5	0.26	0.05	131	25	1310	250
n-butylisocyanate	99.13	115.13	0.04	0.01	0.2	0.05	4	1
Carbon disulfide	76.14	46.3	3	1	156	50	1557	500
Carbon tetrachloride	153.82	76.8	126	20	629	100	4718	750
Chlorine	70.91	-34.05	3	1	9	3	58	20
Chlorine trifluoride	92.50	11.8	0.38	0.1	4	1	38	10
Chloroacetyl chloride	112.94	106	0.5	0.1	5	1	46	10
Chloroform *	119.38	61.7			488	100		
Chloropicrin	164.38	115.1		NA	1	0.2	20	3
Chlorosulfonic acid	116.52	152	2	0.4	10	2.1	30	6.3
Chlorotrifluoroethylene	116.47	-28.22	95	20	476	100	1429	300
Crotonaldehyde	70.09	102.4	6	2	29	10	143	50
Diketene	82.08	127.4	3	1	17	5	168	50
Dimethylamine	45.08	6.88	2	1	184	100	922	500
Epichlorohydrin	92.52	116.4	8	2	76	20	378	100
Ethyl chloride *	64.51	12.27			13192	5000		
Ethylene dichloride *	98.96	83.51			405	100		
Ethylene oxide	44.05	10.5		NA	90	50	901	500
Formaldehyde	30.03	-19.3	1	1	12	10	31	25
Hexachlorobutadiene	260.79	214.2	32	3	107	10	320	30
Hexafluoroacetone	166.02			NA	7	1	339	50
Hydrogen bromide *	80.91	-66.7			17	5		
Hydrogen chloride	36.46	-85.03	4	3	30	20	149	100
Hydrogen cyanide	27.03	25.7		NA	11	10	28	25
Hydrogen fluoride	20.01	19.9	4	5	16	20	41	50
Hydrogen sulfide	34.08	-60.4	0.14	0.1	42	30	139	100
2-isocyanatoethyl methacrylate	155.20	211.2		NA	1	0.1	6	1
Isobutyronitrile	69.11	103.6	28	10	141	50	565	200
Methacrylonitrile *	67.09	90.31			27	10		
Methanol	32.04	64.5	262	200	1310	1000	6551	5000
Methylamine	31.06	-6.32	13	10	127	100	635	500
Methyl chloride	50.49	-24.2		NA	826	400	2065	1000
Methyl iodide	141.94	-66.5	145	25	290	50	726	125
Methyl isocyanate	57.05	38.4	0.058	0.025	1	0.5	12	5
Methyl mercaptan	48.11	5.95	0.01	0.005	49	25	197	100
Perfluoroisobutylene	218.11			NA	1	0.1	3	0.3
Phenol	94.11	181.9	38	10	192	50	770	200

Reference Temperature 25 °C

TABLE 1 (continued)

Material	Molecular Weight	Boiling Point °C	ERPG-1 mg/m³	ERPG-1 PPM	ERPG-2 mg/m³	ERPG-2 PPM	ERPG-3 mg/m³	ERPG-3 PPM
Phosgene	98.92	7.9		NA	1	0.2	4	1
Phosphorous pentoxide	141.94		5	1	25	4	100	17
Propylene oxide *	58.08	34.2			1188	500		
Styrene	104.15	145.2	213	50	1065	250	4259	1000
Sulfur dioxide	64.06	-10	1	0.3	8	3	39	15
Sulfuric acid (Sulfur trioxide)	98.08		2	0.5	10	2.5	30	7.5
Sulfuryl fluoride *	102.06	-55.2			626	150		
Tetrafluoroethylene	100.02	-75.6	818	200	4090	1000	40902	10000
Titanium tetrachloride	189.69	217.45	5	1	20	3	100	13
Toluene diisocyanate *	174.16	252.8			1	0.2		
Trimethylamine	59.11	2.87		0.1	242	100	1209	500
Vinyl acetate	86.09	72.76	18	5	264	75	1760	500
Vinyl chloride *	62.50	-13.8			2556	1000		
Vinylidene chloride *	96.94	31.7			198	50		

NA = Not Appropriate
* – Indicates EEPGs

Reference Temperature 25 °C

To convert ERPG values from PPM to mg/m³, use the following equation:

$$ERPG(mg/m^3) = \frac{ERPG(PPM)\ MW}{24.45}$$

When established ERPG/EEPG values do not exist, the following approaches for deriving substitute values are recommended:

ERPG-2 (in preferred order)

1. Use the workplace exposure guideline (Dow IHG, ACGIH TLV or AIHA WEEL).
   a. Use the STEL or ceiling values if one exists.
   b. Use three times the TWA value.
2. If no workplace guideline exists, contact your industrial hygienist for assistance.

ERPG-3 (in preferred order)

1. LC-50 divided by 30.
2. Use five times the ERPG-2 substitute value

ERPG-1 (in preferred order)

1. Use Odor Threshold value
2. Use ERPG-2 substitute value divided by 10.

Note:

ERPG/EEPG values are updated periodically and the latest values are to be used.

GUIDELINES FOR ESTIMATING THE AMOUNT OF MATERIAL BECOMING AIRBORNE FOLLOWING A RELEASE

This section of the CEI guide provides a description of the method to calculate the airborne quantity. The airborne quantity, as used in this guide, refers to the total quantity of material entering the atmosphere over time, directly as vapor or due to liquid flashing or pool evaporation.

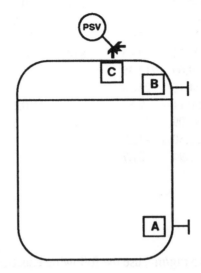

CEI scenarios consider materials to be released as liquid or vapor. For example, the contents of a vessel can escape as a liquid through nozzle A, a vapor through nozzle B or "as calculated" through the relief device attached to nozzle C. Complex calculations that consider two-phase flow from ruptures are not included.

Airborne quantity for vapor releases from nozzle (B) or a pressure relief device (C) is the highest total flow rate calculated given the conditions of the vessel when the release occurs.

Liquid releases require a more complex treatment. As a liquid exits a vessel or pipe as a result of a failure, it can simply run out on the ground forming a pool (see Figure A), partially vaporize forming both a pool and a vapor cloud (see Figure B) or flash to such an extent that all the residual liquid exists as small droplets that are carried away with the vapor (see Figure C).

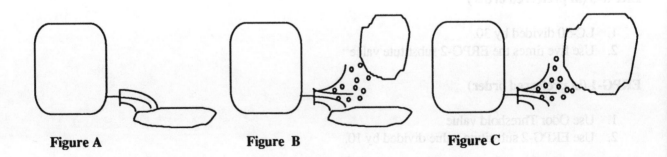

Figure A   **Figure B**   **Figure C**

A simple treatment of these events uses the operating conditions of the process to estimate the behavior of the material after the release.

Liquids reaching the ground form a pool that spreads according to the terrain. If the vessel is surrounded by a dike, the liquid usually flows to the walls of the dike and the pool assumes the area within the dike. In all other cases, the pool is assumed to have an area that is predicted by the amount of liquid that enters the pool. Once a pool is formed, the liquid begins to evaporate from the surface. The vapor from the pool will combine with the vapor from the original flash and be dispersed downwind. This incident is treated by taking a "picture" of the release at a moment in time and then assuming it does not change. (See Figure D)

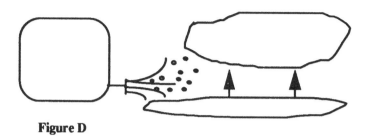

Figure D

The airborne quantity for a liquid spill is determined by what happens to the liquid as it leaves the tank. If the liquid flashes to a high degree, then the airborne quantity is the discharge rate from the vessel. But if the liquid flash is low enough to allow pool formation, the airborne quantity is the gas flow resulting from the flash plus the airborne quantity that evaporates from the pool surface. Finally, as the tendency of the liquid to flash becomes small, the airborne quantity becomes the rate of evaporation from the pool surface.

Figure 2 (page 10) provides a simplified flowchart for calculating airborne quantity. The equations are presented in both SI and US/British units.

FIGURE 2
FLOWCHART FOR CALCULATING AIRBORNE QUANTITY

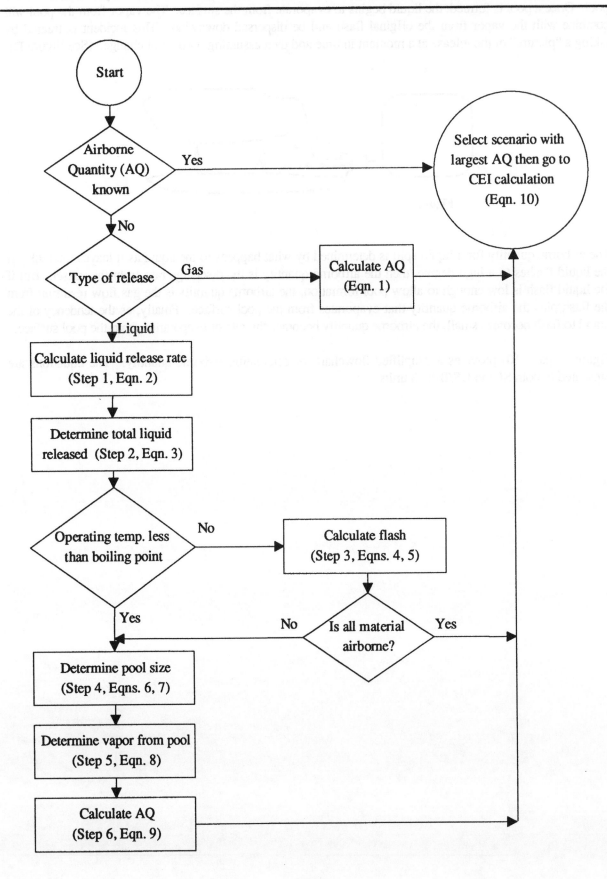

ESTIMATING THE AIRBORNE QUANTITY FOR GAS RELEASES

The following equations, based on the sonic gas flow rate equation, are used to estimate the airborne quantity for a gas release.

SI Units

Airborne Quantity (AQ) $= 4.751 \times 10^{-6} \, D^2 \, P_a \sqrt{\dfrac{MW}{T+273}}$ {kg/sec} (Equation 1A)

where

P_a = absolute pressure = $(P_g + 101.35)$

P_g = gauge pressure (kPa gauge)

MW = molecular weight of the material

T = temperature (°C)

D = diameter of the hole (millimeters)

US/Brit Units

Airborne Quantity (AQ) $= 3.751 \, D^2 \, P_a \sqrt{\dfrac{MW}{T+459}}$ {lb/min} (Equation 1B)

where

P_a = absolute pressure = $(P_g + 14.7)$

P_g = gauge pressure (psig)

MW = molecular weight of the material

T = temperature (°F)

D = diameter of the hole (inches)

ESTIMATING THE AIRBORNE QUANTITY FOR LIQUID RELEASES

The following steps describe a simplified procedure for estimating the airborne quantity for liquid releases.

Step 1: Determine the liquid flow rate being released.

The liquid release rate (L) is given by the following equations:

These equations assume that release from all scenarios will continue for at least five minutes before releases can be stopped. If a five minute release would exceed the total inventory, the release rate is calculated by dividing the total inventory by five minutes.

SI Units

$$L = 9.44 \times 10^{-7} D^2 \rho_1 \sqrt{\frac{1000 P_g}{\rho_1} + 9.8 \Delta h}$$
\{kg/sec\}   (Equation 2A)

where

P_g = gauge pressure (kPa gauge)
(Note: for a tank open to the atmosphere P_g = 0)

ρ_1 = density of the liquid at operating temperature (kg/m³)

Δh = height of the liquid above the release point (meters)

D = diameter of the hole (millimeters)

US/Brit Units

$$L = 2.234 D^2 \rho_1 \sqrt{\frac{144 P_g}{\rho_1} + \Delta h}$$
\{lb/min\}   (Equation 2B)

where

P_g = gauge pressure (psig)
(Note: for a tank open to the atmosphere P_g = 0)

ρ_1 = density of the liquid at operating temperature (lb/ft³)

Δh = height of the liquid above the release point (feet)

D = diameter of the hole (inches)

Step 2: Determine the total liquid released.

The total amount of material contributing to the pool formation must be estimated in order to determine the pool size. If a release is large enough to empty a vessel in less than 15 minutes (including very large releases that occur in less than 5 minutes), the mass of liquid entering the pool is the total inventory of the vessel. For a longer duration continuous release (one lasting more than 15 minutes) the pool is assumed to reach a final size after 15 minutes. In this case, the mass determining the pool size is the release rate times 15 minutes (900 seconds).

The total liquid release (W_T) is the tank inventory (the tank is emptied in less than 15 minutes) or given by:

SI Units

$$W_T = 900 \, L \hspace{4cm} \{kg\} \hspace{2cm} \text{(Equation 3A)}$$

where

L = liquid flow rate (kg/sec)

US/Brit Units

$$W_T = 15 \, L \hspace{4cm} \{lb\} \hspace{2cm} \text{(Equation 3B)}$$

where

L = liquid flow rate (lb/min)

Compare the calculated W_T to the inventory of the system involved in the release. The total liquid assumed to be involved in the release is taken as the smaller of these two values.

W_T = smaller of calculated W_T or system inventory

Step 3: Calculate the fraction flashed.

Compare the operating temperature of the liquid to its normal boiling point. If the temperature is less than the normal boiling point, the flash fraction is zero. Go to Step 4, Equation 6. If the temperature is greater than the normal boiling point, calculate the fraction flashed (F_v).

The fraction of the liquid that will flash (F_v) when released is given by:

$$F_v = \frac{C_p}{H_v}(T_s - T_b) \hspace{4cm} \text{(Equation 4)}$$

where

			SI	US/Brit
T_b	=	normal boiling point of the liquid	°C	°F
T_s	=	operating temperature of the liquid	°C	°F
C_p	=	average heat capacity of the liquid	J/kg/°C	BTU/lb/°F
H_v	=	heat of vaporization of the liquid	J/kg	BTU/lb

The CEI data table contains the ratio of heat capacities to latent heats of vaporization (C_p/H_v) for many chemicals. If a chemical is not listed and the needed information cannot be found, then a value of 0.0044 (SI) or 0.0024 (US/Brit) may be used for the ratio C_p/H_v.

As flashing occurs, some liquid will be entrained as droplets. Some of the droplets are quite small and travel with the vapor while the larger droplets fall to the ground and collect in a pool. As an approximation, the amount of material staying in the vapor is five times the quantity flashed. Therefore, if 20% of the material flashes, the entire stream becomes airborne and there is no pool formed.

The airborne quantity produced by the flash (AQ_f) is given by:

$$AQ_f = 5\,(F_v)\,(L) \qquad \{kg/sec \text{ or } lb/min\} \qquad \text{(Equation 5)}$$

where

L = liquid flow rate (kg/sec or lb/min)

If $F_v \geq 0.2$ then $AQ_f = L$ and no pool is formed. Proceed to Step 6.

Step 4: Determine the pool size.

The total mass of liquid entering the pool (W_p) is given by:

$$W_p = W_T\,(1 - 5F_v) \qquad \{kg \text{ or } lb\} \qquad \text{(Equation 6)}$$

where

W_T = total liquid released (kg or lb)
F_v = fraction flashed

Please note that if none of the material flashes,

$$W_p = W_T \quad \text{(kg or lb)}$$

The size of the pool is approximated by assuming a pool depth of one centimeter. If the spill is in a diked area and of sufficient size, then the pool size is equal to the diked area.

The pool area (A_p) is given by:

SI Units

$$\text{Pool Area }(A_p) = 100\,\frac{W_p}{\rho_l} \qquad \{m^2\} \qquad \text{(Equation 7A)}$$

where

W_p = total mass entering the pool (kg)
ρ_l = density (kg/m^3)

US/Brit Units

$$\text{Pool Area }(A_p) = 30.5\,\frac{W_p}{\rho_l} \qquad \{ft^2\} \qquad \text{(Equation 7B)}$$

W_p = total mass entering the pool (lb)
ρ_l = density (lb/ft^3)

14

If the liquid falls into a diked containment area, then the pool size may be equal to the diked area minus the area taken up by the tank. But, if the spill does not fill the diked area or occurs outside the diked area, use A_p.

Step 5: Determine the airborne quantity evaporated from the pool surface.

Airborne Quantity evaporated from the pool surface (AQ_p) is given by:

SI Units

$$AQ_p = 9.0 \times 10^{-4} \left(A_p^{0.95}\right) \frac{(MW)P_v}{T + 273} \qquad \text{\{kg/sec\}} \qquad \text{(Equation 8A)}$$

where

A_p = pool area (m2)
MW = molecular weight
P_v = vapor pressure of the liquid at the characteristic pool temperature (kPa)
T = characteristic pool temperature (°C) (see Conditions 1 and 2)

US/Brit Units

$$AQ_p = 0.154 \left(A_p^{0.95}\right) \frac{(MW)P_v}{T + 459} \qquad \text{\{lb/min\}} \qquad \text{(Equation 8B)}$$

where

A_p = pool area (ft2)
MW = molecular weight
P_v = vapor pressure of the liquid at the characteristic pool temperature (psi)
T = characteristic pool temperature (°F) (see Conditions 1 and 2)

Condition 1
If the liquid is at or above ambient temperature but below its normal boiling point, the characteristic pool temperature is equal to the operating temperature.

Condition 2
If the liquid is at or above its normal boiling point, the characteristic pool temperature is the normal boiling point of the liquid. The normal boiling point is the boiling point of the liquid at atmospheric pressure.

Step 6: Calculate the total airborne quantity.

The total airborne quantity (AQ) is calculated by:

$$AQ = AQ_f + AQ_p \qquad \text{\{kg/sec or lb/min\}} \qquad \text{(Equation 9)}$$

where

AQ_f = airborne quantity resulting from the flash (kg/sec or lb/min)
AQ_p = airborne quantity evaporating from the pool surface (kg/sec or lb/min)

If the total Airborne Quantity (AQ) is greater than the liquid flow rate (L), set AQ = L.

Chemical Exposure Index

All CEI calculations assume a windspeed of 5 m/sec (11.2 miles/hour) and neutral weather conditions.

The Chemical Exposure Index (CEI) is given by:

SI Units

$$CEI = 655.1\sqrt{\frac{AQ}{ERPG\text{-}2}}$$ (Equation 10A)

where

AQ = airborne quantity (kg/sec)
ERPG-2 = value (mg/m^3)

US/Brit Units

$$CEI = 281.8\sqrt{\frac{AQ}{ERPG\text{-}2(MW)}}$$ (Equation 10B)

where

AQ = airborne quantity (lb/min)
ERPG-2 = value (PPM)
MW = molecular weight

If the CEI calculated value is greater than 1000, set CEI = 1000.

Hazard Distance

The Hazard Distance (HD) is the distance to the ERPG-1, -2 or -3 concentration and is derived from the following equation:

SI Units

$$HD = 6551\sqrt{\frac{AQ}{ERPG}}$$
{meters}   (Equation 11A)

where

AQ = airborne quantity (kg/sec)
ERPG = ERPG-1, ERPG-2 or ERPG-3 (mg/m^3)

US/Brits Units

$$HD = 9243\sqrt{\frac{AQ}{ERPG(MW)}}$$
{feet}   (Equation 11B)

where

AQ = airborne quantity (lb/min)
ERPG = ERPG-1, ERPG-2 or ERPG-3 (PPM)
MW = molecular weight

If HD is greater than 10,000 meters (32,800 feet), set HD = 10,000 meters (32,800 feet).

CEI VS. AIRBORNE QUANTITY – SI Units

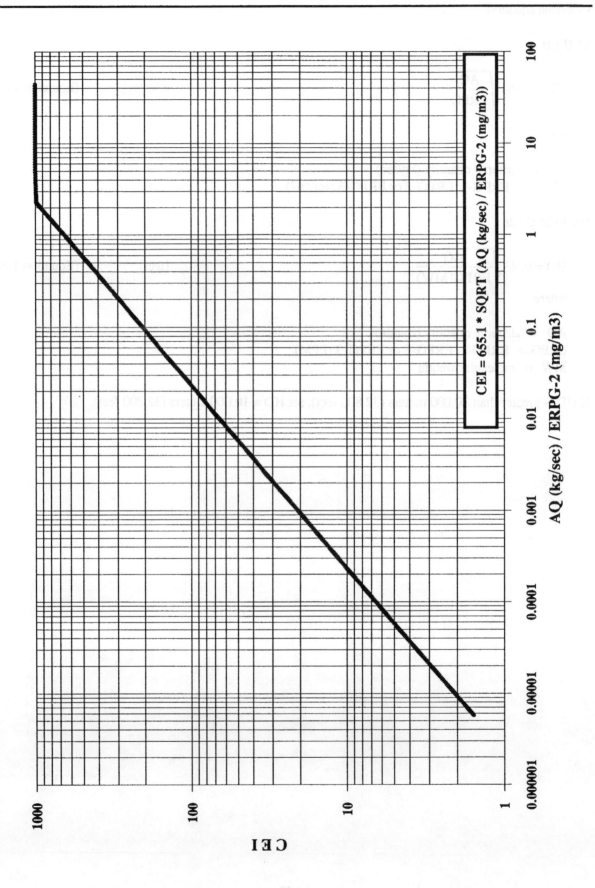

CEI = 655.1 * SQRT (AQ (kg/sec) / ERPG-2 (mg/m3))

AQ (kg/sec) / ERPG-2 (mg/m3)

C E I

CEI VS. AIRBORNE QUANTITY – US/BRIT Units

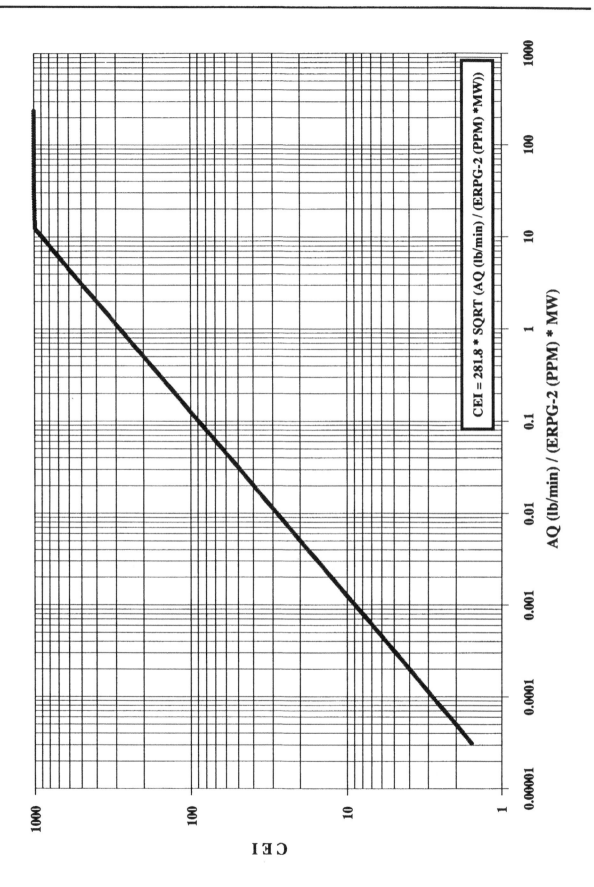

CEI = 281.8 * SQRT (AQ (lb/min) / (ERPG-2 (PPM) *MW))

AQ (lb/min) / (ERPG-2 (PPM) * MW)

C E I

CHEMICAL EXPOSURE INDEX SUMMARY

Plant _____ Location _____

Chemical _____ Total Quantity In Plant _____

Largest Single Containment _____

Pressure Of Containment _____ Temperature Of Containment _____

1. Scenario Being Evaluated _____

2. Airborne Release Rate from Scenario   _____ kg/sec
        _____ lb/min

3. **Chemical Exposure Index**   _____

4.

	Concentration			Hazard Distance	
	mg/m^3	PPM		meters	feet
ERPG-1/EEPG-1	_____	_____		_____	_____
ERPG-2/EEPG-2	_____	_____		_____	_____
ERPG-3/EEPG-3	_____	_____		_____	_____

5. Distances to:

	meters	feet
Public (generally considered property line)	_____	_____
Other in-company facility	_____	_____
Non-company plant or business	_____	_____

6. The CEI and the Hazard Distance establish the level of review needed.

7. If further review is required, complete Containment and Mitigation Checklist (*Chemical Exposure Index Guide*, 2nd Edition – Appendix 2, page 26) and prepare Review Package.

8. List any sights, odors or sounds that might come from your facility and cause public concern or inquiries (e.g., smoke, large relief valves, odors below hazardous levels such as mercaptans or amines, etc.)

_____

_____

_____

Prepared by: _____

Reviewed by: _____

Date

Plant Superintendent or Manager   _____ _____

Site Review Representative   _____ _____

Additional Management Review
   (if required)   _____ _____

TABLE 2A

PHYSICAL PROPERTY TABLE FOR CEI (SI UNITS)

Chemical	Molecular Weight	Boiling Point °C	Vapor Pressure kPa	Liquid Density @ 25 °C kg/m³	Liquid Density @ BP kg/m³	Gas Density @ 25 °C kg/m³	Ratio of C_p/H_v 1/°C
Acrolein	56.06	52.5	35.866	834.4			
Acrylic acid	72.06	141.4	0.539	1046.0			
Acrylonitrile	53.06	77.2	13.900	801.0			
Allyl chloride	76.53	44.8	48.480	931.4			
Ammonia	17.03	-33.4	1002.800	602.3	684.0	7.79	4.01E-03
Benzene	78.11	80.1	12.690	869.8			
Bromine	159.81	58.7	28.375	3105.0			
Butadiene	54.09	-4.4	281.090	614.9	651.0	6.69	5.92E-03
Carbon disulfide	76.14	46.3	48.120	1256.0			
Carbon monoxide	29.01	-191.5	2807.000				
Carbon tetrachloride	153.82	76.8	15.162	1585.0			
Chlorine	70.91	-34.0	778.340	1399.0	1562.0	25.07	3.87E-03
Chloroacetyl chloride	112.94	106.0	3.330	1412.0			
Chloroform	119.38	61.7	26.175	1480.0			
Chloropicrin	164.38	112.0	3.356	1648.0			
Chlorotrifluoroethylene	116.47	-28.2	641.260	1290.0	1472.0	35.13	7.98E-03
Crotonaldehyde	70.09	102.4	4.787	848.9			
Dimethylamine	45.08	6.9	205.460	649.7	671.0	3.96	4.89E-03
Epichlorohydrin	92.52	116.4	2.279	1175.0			
Ethyl chloride	64.51	12.3	159.950	892.1	910.0	4.40	4.31E-03
Ethylene dichloride	98.96	83.5	10.590	1246.0			
Ethylene oxide	44.05	10.5	174.010	866.8	887.0	3.25	3.65E-03
Hydrogen bromide	80.91	-66.7	2358.800	1762.0	2199.0	95.46	5.66E-03
Hydrogen chloride	36.46	-85.0	4773.100	805.2	1188.0	113.0	9.81E-03
Hydrogen cyanide	27.03	25.7	98.780	679.6	679.0	1.14	2.83E-03
Hydrogen fluoride	20.01	19.6	122.740	981.5	991.0	1.15	4.24E-03
Hydrogen sulfide	34.08	-60.3	2020.800	767.3	929.0	34.30	5.26E-03
Methacrylonitrile	67.09	90.3	9.477	794.9			
Methanol	32.04	64.5	16.950	786.0			
Methylamine	31.06	-6.3	348.440	655.2	694.0	4.66	3.92E-03
Methyl chloride	50.49	-24.1	576.540	915.7	1014.0	13.12	4.19E-03
Methyl mercaptan	48.11	6.0	201.820	858.6	884.0	4.12	3.87E-03
Phenol	94.11	181.9	0.055	1070.0			
Phosgene	98.92	7.5	189.900	1360.0	1403.0	7.96	4.32E-03
Propylene oxide	58.08	34.2	71.670	823.2			
Styrene	104.15	145.2	0.841	901.6			
Sulfuryl fluoride	102.06	-55.2	1747.100	1318.0	1702.0	97.38	9.57E-03
Sulfur dioxide	64.06	-10.0	392.850	1353.0	1444.0	10.86	3.91E-03
Sulfur trioxide	80.06	44.4	35.688	1904.0			
Toluene diisocyanate	174.16	252.9	0.002	1211.0			
Trimethylamine	59.11	2.9	221.160	624.8	653.0	5.68	6.15E-03
Vinyl acetate	86.09	72.8	15.280	924.7			
Vinyl chloride	62.50	-13.8	395.140	902.1	972.0	10.90	3.88E-03
Vinylidene chloride	96.94	31.7	79.517	1203.0			

TABLE 2B

PHYSICAL PROPERTY TABLE FOR CEI (US/BRIT UNITS)

Chemical	Molecular Weight	Boiling Point °F	Vapor Pressure psia	Liquid Density @ 77 °F lb/ft³	@ BP lb/ft³	Gas Density @ 77 °F lb/ft³	Ratio of C_p/H_v 1/°F
Acrolein	56.06	126.5	5.201	52.1			
Acrylic acid	72.06	286.5	0.078	65.3			
Acrylonitrile	53.06	171.0	2.016	50.0			
Allyl chloride	76.53	112.6	7.030	58.1			
Ammonia	17.03	-28.0	145.408	37.6	42.7	0.49	2.23E-03
Benzene	78.11	176.2	1.840	54.3			
Bromine	159.81	137.7	4.114	193.8			
Butadiene	54.09	24.0	40.759	38.4	40.6	0.42	3.29E-03
Carbon disulfide	76.14	115.3	6.978	78.4			
Carbon monoxide	29.01	-312.7	407.022				
Carbon tetrachloride	153.82	170.2	2.199	99.0			
Chlorine	70.91	-29.2	112.861	87.3	97.5	1.57	2.15E-03
Chloroacetyl chloride	112.94	222.8	0.483	88.2			
Chloroform	119.38	143.1	3.795	92.4			
Chloropicrin	164.38	233.5	0.487	102.9			
Chlorotrifluoroethylene	116.47	-18.8	92.984	80.5	91.9	2.19	4.43E-03
Crotonaldehyde	70.09	216.3	0.694	53.0			
Dimethylamine	45.08	44.4	29.792	40.6	41.9	0.25	2.72E-03
Epichlorohydrin	92.52	241.5	0.330	73.4			
Ethyl chloride	64.51	54.1	23.193	55.7	56.8	0.27	2.40E-03
Ethylene dichloride	98.96	182.3	1.536	77.8			
Ethylene oxide	44.05	50.9	25.232	54.1	55.4	0.20	2.03E-03
Hydrogen bromide	80.91	-88.1	342.032	110.0	137.3	5.96	3.14E-03
Hydrogen chloride	36.46	-121.1	692.111	50.3	74.2	7.05	5.45E-03
Hydrogen cyanide	27.03	78.3	14.323	42.4	42.4	0.07	1.57E-03
Hydrogen fluoride	20.01	67.3	17.798	61.3	61.9	0.07	2.36E-03
Hydrogen sulfide	34.08	-76.5	293.021	47.9	58.0	2.14	2.92E-03
Methacrylonitrile	67.09	194.6	1.374	49.6			
Methanol	32.04	148.1	2.458	49.1			
Methylamine	31.06	20.6	50.525	40.9	43.3	0.29	2.18E-03
Methyl chloride	50.49	-11.4	83.600	57.2	63.3	0.82	2.33E-03
Methyl mercaptan	48.11	42.7	29.264	53.6	55.2	0.26	2.15E-03
Phenol	94.11	359.4	0.008	66.8			
Phosgene	98.92	45.5	27.536	84.9	87.6	0.50	2.40E-03
Propylene oxide	58.08	93.6	10.392	51.4			
Styrene	104.15	293.4	0.122	56.3			
Sulfuryl fluoride	102.06	-67.4	253.334	82.3	106.3	6.08	5.32E-03
Sulfur dioxide	64.06	14.0	56.964	84.5	90.1	0.68	2.17E-03
Sulfur trioxide	80.06	111.9	5.175	118.9			
Toluene diisocyanate	174.16	487.2	0.000	75.6			
Trimethylamine	59.11	37.2	32.069	39.0	40.8	0.35	3.41E-03
Vinyl acetate	86.09	163.0	2.216	57.7			
Vinyl chloride	62.50	7.1	57.296	56.3	60.7	0.68	2.16E-03
Vinylidene chloride	96.94	89.1	11.530	75.1			

APPENDIX 1

CHEMICAL EXPOSURE INDEX REVIEW PROCESS

The following is a recommended procedure for the CEI review process.

When a plant or facility requires a further review, they should notify the site Loss Prevention contact and agree to an appropriate review schedule.

The review process should have three elements:
1. A Pre-Review Working Session
2. The Chemical Exposure Index Review Package
3. The Formal Review

Pre-Review Working Session

A one to two hour working session at least two weeks prior to the formal review is strongly recommended.

Purpose: This preliminary working session would let the actual review focus on what can and will be done to eliminate, reduce and/or mitigate potential releases.

This working session should have the following suggested agenda:

1. Review each chemical scenario, lines of defense, mitigation measures and plans for improvement with review team.
2. Discuss any past releases of acutely toxic material.
3. Interview an operator in the plant.
   This interview should focus on the operator's awareness of the Chemical Exposure Index scenarios, use of emergency procedures and specific personal concerns related to the potential release of toxic chemicals.
4. Conduct a drill of a hypothetical CEI release scenario. This drill should be conducted by plant personnel with CEI review team members present and should be designed to evaluate the response to an emergency situation involving a major release of one of the chemicals with the largest CEI.
5. Inspect equipment and piping related to each scenario with the largest CEI.
   This inspection will consider the condition of pipe and equipment, location of valves and mitigating devices, etc.
6. Review documentation pertaining to each CEI scenario.
   The following documents are suggested as a minimum:
   - Past hypothetical exercise reports
   - Area monitoring system records
   - Maintenance checklists
   - Operating Discipline including: loading/unloading procedures, shutdown for releases, mitigation procedures, spill reporting procedure, etc.
   Completeness and quality will be used as a criterion for evaluating the documentation.

Written reports of the interview, drill, site inspection and documentation review should be prepared.

Chemical Exposure Review Package

At least one week before the review, the following information should be sent to each member of the review team:

1. Chemical Exposure Index Summary sheets for all chemicals and scenarios calculated.
2. A simplified process flowsheet for areas with chemicals having the largest CEIs.
   Include: a) all vessels including description, designation, (i.e., storage tank VT-100) size and normal contents.
   b) associated piping including size, approximate length between equipment and any automated block valves.
3. A description of the scenario for each chemical that results in the largest CEIs.
   Include: a) description of any major changes made to this system since the last audit.
   b) description of the lines of defense or actions to be taken if each scenario occurs.
   c) description of any mitigation methods, including size of spill areas or dike areas.
4. An area map showing three circles that represent the hazard distance for ERPG-1, -2 and -3 concentrations.
5. A plot plan showing the location of gas monitors, spill detectors or other devices used to detect releases. Also include the type of device used to detect the release and the detection level for each type of device.
6. A completed containment and compliance checklist.
7. A list of the recommendations made in the previous review and the status of each recommendation.
8. A written report for the hypothetical exercise based on one of the Chemical Exposure Index scenarios. Include a description of the hypothetical release, a chronological list of actions taken and any recommendations for improvement. Include time it took to isolate the release.
   Note: If the hypothetical exercise is to be done during the review, please include the written description of the exercise scenario in the package.
9. Written results of the employee interview.
10. A written report of the in-plant inspection of each Chemical Exposure Index scenario source site.

If a pre-review working session was not held, these last two items must be provided during the formal review.

Formal Review

The following four activities should be done prior to formal presentation if a pre-review working session was not held:

1. Hypothetical exercise
2. Employee interview
3. Site inspection of each scenario site
4. Documentation review

Suggested Review Agenda

1. Summarize the status of the recommendations made in the previous review.
2. Discuss incidents that have occurred involving chemicals being reviewed.
   Include Spill History – incident reports and plant follow-up reports.
3. Summarize the results of CEI calculations. Include hazard distances for ERPG-1, -2, -3.

4. Review the release scenarios for the largest CEIs. Include plot plan showing hazard distances. Address the following items for each scenario:
   a. Have there been any major changes or additions to this system since the last audit/assessment?
   b. Discuss your lines of defense that would prevent this scenario release from occurring.
   c. Discuss the mitigation procedures for each scenario.
5. Discuss results of hypothetical exercise held prior to audit.
6. Discuss results of employee interview held prior to audit.
7. Review results of site inspection of each CEI scenario sources site.
8. Discuss results of documentation review.
9. Discuss your plant's CEI related concerns generated from the audit/assessment preparation.
10. Discuss your plans for improving your operations from a CEI standpoint.
11. Are there any other specific concerns which should be addressed but were not identified in the CEI review?

A written response including action to be taken, person assigned, anticipated completion date should be reported to the Line Management and the Review Chairman within 45 days and maintained with your Chemical Exposure Index calculations.

APPENDIX 2

CONTAINMENT AND MITIGATION CHECKLIST

This checklist is a Process Hazard Analysis tool for evaluating a facility's mitigation features to prevent, detect or control potential releases of acute toxic substances.

The management of any plant or facility requiring review should complete this checklist as part of the preparation for a Chemical Exposure Index review. Please check those that are complete, operational or in compliance with known company rules, guidelines or requirements. Any item that cannot be checked should be marked with the percentage of completion. Be prepared to discuss plans for any item not completed.

Complete (✓)	Risk Reducing Factors
_____	1. All pressure vessel and relief device systems properly registered and inspection up to date and documentation complete. (No expansion joints or glass devices.)
_____	2. All hoses inspected and tested regularly.
_____	3. All operational controls and systems designed and routinely tested to "fail-safe."
_____	4. Critical Instrument Program up to date (e.g., redundant high level and temperature alarms, shutdowns, etc.)
_____	5. Operating Discipline complete and up to date.
_____	6. Vapor Detectors properly placed and tested regularly.
_____	7. Appropriate engineering specifications properly applied (e.g., lethal service, welded fittings, etc.)
_____	8. Are relief vents on toxic containers designed to minimize atmospheric emissions? How? (circle) Scrubber, Flare or _____
_____	9. Failure analysis and nondestructive testing carried out where needed (e.g., X-ray, vibration analysis or monitoring, acoustical emission, piping flexibility – hot and cold).
_____	10. Physical barriers in place (for traffic, cranes, etc.)
_____	11. Designed for excess pressure, if needed (e.g., pipelines in certain areas, tank cars, trucks, etc.).
_____	12. All personnel properly trained to understand hazards and emergency responses.
_____	13. Emergency Procedures (relating to this exposure potential) in place and annual drill held.
_____	14. Safety Rules and Safety Standards regularly reviewed and enforced.
_____	15. Loss Prevention Principles and Minimum Requirements appropriately applied.
_____	16. Technology Center Guidelines appropriately incorporated.
_____	17. Reactive Chemical Review complete and up to date.
_____	18. Loss Prevention Audit complete and up to date.
_____	19. Technology Center Audit complete and up to date.
_____	20. All new operations and modifications underwent safety pre-startup audit.
_____	21. Management of Change procedures written and utilized.

Completed by: _____ **Date:** _____

Reviewed by: _____

26

APPENDIX 3

EXAMPLE CEI CALCULATIONS

CHLORINE VAPOR RELEASE

The 3/4 inch vapor connection on a 1 ton chlorine cylinder stored at ambient temperature (30 °C or 86 °F) has broken.

Needed information:

	SI	US/Brit
Pressure inside the cylinder, P_g	788.1 kPa gauge	114.3 psig
Absolute Pressure, Pa	889.5 kPa	129.0 psia
Molecular Weight, MW	70.91	70.91
Storage Temperature, T	30 °C	86 °F
Diameter of Hole, D	19 mm	0.75 in

Determine Airborne Quantity.

SI Units (Equation 1A)

$$AQ = 4.751 \times 10^{-6} \, D^2 \, Pa \sqrt{\frac{MW}{(T+273)}}$$

$$AQ = 4.751 \times 10^{-6} (19)^2 (889.5) \sqrt{\frac{70.91}{(30+273)}}$$

$$AQ = 0.74 \, kg/sec$$

US/Brit Units (Equation 1B)

$$AQ = 3.751 \, D^2 \, Pa \sqrt{\frac{MW}{(T+459)}}$$

$$AQ = 3.751 (.75)^2 (129.0) \sqrt{\frac{70.91}{(86+459)}}$$

$$AQ = 98.2 \, lb/min$$

Calculate the CEI.

SI Units (Equation 10A)

where ERPG-2 = 9 mg/m^3

$$CEI = 655.1 \sqrt{\frac{AQ}{ERPG\text{-}2}}$$

$$CEI = 655.1 \sqrt{\frac{0.74}{9.0}}$$

$$CEI = 188$$

US/Brit Units (Equation 10B)

where ERPG-2 = 3 PPM

$$CEI = 281.8 \sqrt{\frac{AQ}{(ERPG\text{-}2)(MW)}}$$

$$CEI = 281.8 \sqrt{\frac{98.2}{(3.0)(70.91)}}$$

$$CEI = 191$$

Differences in CEI values result from rounding the ERPG values when converting between PPM and mg/m^3.

Calculate Hazard Distances.

SI Units (Equation 11A)	*US/Brit Units (Equation 11B)*

For ERPG-2 = 9 mg/m³

$$HD = 6551\sqrt{\dfrac{AQ}{ERPG}}$$

$$HD = 6551\sqrt{\dfrac{0.74}{9}}$$

HD = 1,878 meters

For ERPG-1 = 3 mg/m³

$$HD = 6551\sqrt{\dfrac{0.74}{3}}$$

HD = 3,254 meters

For ERPG-3 = 58 mg/m³

$$HD = 6551\sqrt{\dfrac{0.74}{58}}$$

HD = 740 meters

For ERPG-2 = 3 PPM

$$HD = 9243\sqrt{\dfrac{AQ}{ERPG(MW)}}$$

$$HD = 9243\sqrt{\dfrac{98.2}{3(70.91)}}$$

HD = 6,280 feet

For ERPG-1 = 1 PPM

$$HD = 9243\sqrt{\dfrac{98.2}{1(70.91)}}$$

HD = 10,878 feet

For ERPG-3 = 20 PPM

$$HD = 9243\sqrt{\dfrac{98.2}{20(70.91)}}$$

HD = 2,432 feet

Differences in HD values result from rounding the ERPG values when converting between PPM and mg/m³.

28

AMMONIA LIQUID RELEASE

Ammonia is stored in a 12 ft diameter by 72 ft long horizontal vessel under its own vapor pressure at ambient temperature (30 °C or 86 °F). The largest liquid line out of the vessel is 2 inch diameter (50.8 mm).

Needed information:

	SI	US/Brit
Pressure inside vessel, P_g	1064 kPa gauge	154.5 psig
Temperature inside vessel, T	30 °C	86 °F
Normal boiling point	-33.4 °C	-28 °F
Liquid density in vessel, ρ_l	594.5 kg/m^3	37.1 lb/ft^3
Ratio C_p/H_v	4.01 E-03	2.23 E-03
Height of liquid in tank, Δh	3.66 m	12 ft
Diameter of hole, D	50.8 mm	2.0 in
Molecular weight, MW	17.03	17.03

Estimate liquid released.

SI Units (Equation 2A)

$$L = 9.44 \times 10^{-7} D^2 \rho_l \sqrt{\frac{1000 P_g}{\rho_l} + 9.8\,\Delta h}$$

$$L = 9.44 \times 10^{-7} (50.8)^2 (594.5) \sqrt{\frac{1000(1064)}{594.5} + 9.8\,(3.66)}$$

$$L = 61.9 \text{ kg / sec}$$

US/Brit Units (Equation 2B)

$$L = 2.234 D^2 \rho_l \sqrt{\frac{144 P_g}{\rho_l} + \Delta h}$$

$$L = 2.234 (2.0)^2 (37.1) \sqrt{\frac{144(154.5)}{37.1} + 12.0}$$

$$L = 8,200 \text{ lb / min}$$

Estimate flash fraction.

SI Units (Equation 4A)

$$F_v = \frac{C_p}{H_v}(T_s - T_b)$$

$$F_v = 0.00401(30 - (-33.4))$$

$$F_v = 0.254$$

Since $F_v > 0.2 \quad AQ = L$

$$AQ = 61.9 \text{ kg/sec}$$

US/Brit Units (Equation 4B)

$$F_v = \frac{C_p}{H_v}(T_s - T_b)$$

$$F_v = 0.00223(86 - (-28))$$

$$F_v = 0.254$$

Since $F_v > 0.2 \quad AQ = L$

$$AQ = 8,200 \text{ lb/min}$$

Calculate CEI.

SI Units (Equation 10A)

where ERPG-2 $= 139$ mg/m^3

$$CEI = 655.1 \sqrt{\frac{AQ}{ERPG\text{-}2}}$$

$$CEI = 655.1 \sqrt{\frac{61.9}{139}}$$

$$CEI = 437$$

US/Brit Units (Equation 10B)

where ERPG-2 $= 200$ PPM

$$CEI = 281.8 \sqrt{\frac{AQ}{(ERPG\text{-}2)(MW)}}$$

$$CEI = 281.8 \sqrt{\frac{8200}{200(17.03)}}$$

$$CEI = 437$$

Calculate the Hazard Distances.

SI Units (Equation 11A)

For ERPG-2 $= 139$ mg/m^3

$$HD = 6551 \sqrt{\frac{AQ}{ERPG}}$$

$$HD = 6551 \sqrt{\frac{61.9}{139}}$$

$$HD = 4,372 \text{ meters}$$

For ERPG-1 $= 17$ mg/m^3

$$HD = 6551 \sqrt{\frac{61.9}{17}}$$

$$HD = 12,500 \text{ meters}$$

For ERPG-3 $= 696$ mg/m^3

$$HD = 6551 \sqrt{\frac{61.9}{696}}$$

$$HD = 1,953 \text{ meters}$$

US/Brit Units (Equation 11B)

For ERPG-2 $= 200$ PPM

$$HD = 9243 \sqrt{\frac{AQ}{ERPG(MW)}}$$

$$HD = 9243 \sqrt{\frac{8200}{200(17.03)}}$$

$$HD = 14,342 \text{ feet}$$

For ERPG-1 $= 25$ PPM

$$HD = 9243 \sqrt{\frac{8200}{25(17.03)}}$$

$$HD = 40,564 \text{ feet}$$

For ERPG-3 $= 1000$ PPM

$$HD = 9243 \sqrt{\frac{8200}{1000(17.03)}}$$

$$HD = 6,414 \text{ feet}$$

STYRENE LIQUID RELEASE

Styrene is stored in a 40 ft x 40 ft API tank at ambient temperature (25 °C or 77 °F). The tank has a closed vent system but is essentially at ambient pressure. The outlet is a 6-inch Schedule 40 nozzle.

Needed information:

	SI	US/Brit
Pressure inside the tank, P_g	0.0 kPa	0.0 psig
Temperature inside the tank, T	25 °C	77 °F
Normal boiling point	145.2 °C	293.4 °F
Vapor pressure, ambient temperature	0.841 kPa	0.122 psi
Liquid density, ρ_l	901.6 kg/m3	56.3 lb/ft3
Height of liquid, Δh	12.2 m	40.0 ft
Molecular weight, MW	104.15	104.15

Scenario selection — For greater than 4-inch diameter, use 20% of the cross sectional area (CSA).

For 6-inch Schedule 40, CSA = 28.89 in2

$$0.20(28.89) = 5.78 \text{ in}^2$$

$$D = \sqrt{\frac{4}{\pi}A} = \sqrt{\frac{4}{\pi}5.78} = 2.71 \text{ in or } 68.9 \text{ mm}$$

Estimate liquid released.

SI Units (Equation 2A)

$$L = 9.44 \times 10^{-7} D^2 \rho_l \sqrt{\frac{1000 P_g}{\rho_l} + 9.8\,\Delta h}$$

$$L = 9.44 \times 10^{-7} (68.9)^2 (901.6) \sqrt{\frac{1000(0)}{901.6} + 9.8\,(12.2)}$$

$$L = 44.2 \text{ kg / sec}$$

US/Brit Units (Equation 2B)

$$L = 2.234 D^2 \rho_l \sqrt{\frac{144 P_g}{\rho_l} + \Delta h}$$

$$L = 2.234 (2.71)^2 (56.3) \sqrt{\frac{144(0)}{56.3} + 40.0}$$

$$L = 5,842 \text{ lb / min}$$

Compare operating temperature to normal boiling point:

25 °C < 145 °C

77 °F < 293.4 °F

Therefore, Flash Fraction = 0

Estimate pool size.

SI Units (Equation 3A)

$$W_T = 900(L)$$

$$W_T = 39,800 \text{ kg} = W_p$$

US/Brit Units (Equation 3B)

$$W_T = 15(L)$$

$$W_T = 87,600 \text{ lb} = W_p$$

SI Units (Equation 7A)

$$A_p = 100 \frac{W_p}{\rho_l}$$

$$A_p = 100 \frac{39800}{901.6}$$

$$A_p = 4,410 \, m^2$$

Assume no dike.

SI Units (Equation 8A)

$$AQ_p = 9.0 \times 10^{-4} \left(A_p^{0.95} \right) \frac{(MW) P_v}{T + 273}$$

Characteristic pool temperature equals ambient

$$AQ_p = 9.0 \times 10^{-4} \left(4410^{0.95} \right) \frac{104.15(0.841)}{25 + 273}$$

$$AQ_p = 0.767 \, kg / sec$$

SI Units (Equation 9A)

$$AQ = AQ_f + AQ_p$$

$$AQ = 0 + 0.767$$

$$AQ = 0.729 \, kg / sec$$

Calculate CEI.

SI Units (Equation 10A)

where ERPG-2 = 1065 mg/m³

$$CEI = 655.1 \sqrt{\frac{AQ}{ERPG\text{-}2}}$$

$$CEI = 655.1 \sqrt{\frac{0.767}{1065}}$$

$$CEI = 18$$

US/Brit Units (Equation 7B)

$$A_p = 30.5 \frac{W_p}{\rho_l}$$

$$A_p = 30.5 \frac{87600}{56.3}$$

$$A_p = 47,460 \, ft^2$$

US/Brit Units (Equation 8B)

$$AQ_p = 0.154 \left(A_p^{0.95} \right) \frac{(MW) P_v}{T + 459}$$

$$AQ_p = 0.154 \left(47460^{0.95} \right) \frac{104.15(0.122)}{77 + 459}$$

$$AQ_p = 101 \, lb / min$$

US/Brit Units (Equation 9B)

$$AQ = AQ_f + AQ_p$$

$$AQ = 0 + 101$$

$$AQ = 101 \, lb / min$$

US/Brit Units (Equation 10B)

where ERPG-2 = 250 PPM

$$CEI = 281.8 \sqrt{\frac{AQ}{(ERPG\text{-}2)(MW)}}$$

$$CEI = 281.8 \sqrt{\frac{101}{250(104.15)}}$$

$$CEI = 18$$

32

Calculate the Hazard Distances.

<table>
<tr><td>

SI Units (Equation 11A)

For ERPG-2 $= 1065$ mg/m^3

$$HD = 6551\sqrt{\dfrac{AQ}{ERPG}}$$

$$HD = 6551\sqrt{\dfrac{0.767}{1065}}$$

HD = 176 meters

For ERPG-1 $= 213$ mg/m^3

$$HD = 6551\sqrt{\dfrac{0.767}{213}}$$

HD = 393 meters

For ERPG-3 $= 4259$ mg/m^3

$$HD = 6551\sqrt{\dfrac{0.767}{4259}}$$

HD = 87.9 meters

</td><td>

US/Brit Units (Equation 11B)

For ERPG-2 $= 250$ PPM

$$HD = 9243\sqrt{\dfrac{AQ}{ERPG(MW)}}$$

$$HD = 9243\sqrt{\dfrac{101}{250(104.15)}}$$

HD = 576 feet

For ERPG-1 $= 50$ PPM

$$HD = 9243\sqrt{\dfrac{101}{50(104.15)}}$$

HD = 1,287 feet

For ERPG-3 $= 1000$ PPM

$$HD = 9243\sqrt{\dfrac{101}{1000(104.15)}}$$

HD = 288 feet

</td></tr>
</table>

CHLORINE LIQUID RELEASE

Chlorine is stored in a sphere at 5 °C (41 °F). A 2-inch nozzle fails on the bottom of the vessel allowing liquid to escape.

Needed information:

	SI	US/Brits
Pressure inside the cylinder, P_g	332 kPa gauge	48.2 psig
Molecular weight, MW	70.91	70.91
Storage temperature, T	5 °C	41 °F
Liquid density, ρ_1	1458 kg/m^3	91.01 lb/ft^2
Height of liquid in the sphere, Δh	6 m	19.7 ft
Diameter of hole, D	50.8 mm	2 in
Capacity of sphere	1.134 x 10^6 kg	2.5 x 10^6 lb

Estimate liquid released.

SI Units (Equation 2A)

$$L = 9.44 \times 10^{-7} D^2 \rho_1 \sqrt{\frac{1000 P_g}{\rho_1} + 9.8\,\Delta h}$$

$$L = 9.44 \times 10^{-7} (50.8)^2 (1458) \sqrt{\frac{1000(332)}{1458} + 9.8\,(6)}$$

$$L = 60.1 \text{ kg / sec}$$

US/Brit Units (Equation 2B)

$$L = 2.234 D^2 \rho_1 \sqrt{\frac{144 P_g}{\rho_1} + \Delta h}$$

$$L = 2.234 (2)^2 91.01 \sqrt{\frac{144(48.2)}{91.01} + 19.7}$$

$$L = 7,967 \text{ lb / min}$$

Determine the total liquid released.

For 15 minutes (900 seconds), the total liquid leaving the tank is:

$$W_T = 900(60.1) = 54,090 \text{ kg}$$

$$W_T = 15(7,967) = 119,505 \text{ lb}$$

The capacity of the tank when full is 1.134 x 10^6 kg. Since L_T = 54090 kg is less than the capacity of the tank.

$$W_T = 54,090 \text{ kg}$$

$$W_T = 119,505 \text{ lb}$$

Calculate the flash fraction.

Needed information:

Normal boiling point temperature = -34 °C   Normal boiling point temperature = -29.2 °F
Heat of vaporization = 275,030 J/kg
Heat capacity of liquid (at average temperature) = 943.8 J/kg/°C

A technically correct solution for evaluating the flash fraction requires the heat capacity (C_p) to be evaluated at the average temperature (storage and boiling point) and the heat of vaporization at the boiling point. For example:

$$C_p \ (@ -15 \ °C \ or \ 5 \ °F) = 943.8 \ J/kg/°C = 0.2254 \ BTU/lb/°F$$

and

$$H_v \ (BP) = 285{,}457 \ J/kg = 122.72 \ BTU/lb$$

SI Units (Equation 4A)	*US/Brit Units (Equation 4B)*
$$F_v = \frac{C_p}{H_v}(T_s - T_b)$$	$$F_v = \frac{C_p}{H_v}(T_s - T_b)$$
$$F_v = \frac{943.8}{285{,}457}(5-(-34.0))$$	$$F_v = \frac{0.2254}{122.72}(41-(-29.2))$$
$$F_v = 0.129$$	$$F_v = 0.129$$

Calculate vapor source strength from the flash.

$$AQ_f = 5(F_v)(L) = 5(0.129)(60.1) = 38.8 \ kg/sec \qquad \text{(SI)}$$

$$AQ_f = 5(F_v)(L) = 5(0.129)(7967) = 5{,}139 \ lb/min \qquad \text{(US/Brit)}$$

Calculate the total liquid entering the pool.

$$W_p = W_T(1-5F_v) = 54{,}090(1-(5)(0.129)) = 19{,}202 \ kg \qquad \text{(SI)}$$

$$W_p = W_T(1-5F_v) = 119{,}505(1-(5)(0.129)) = 42{,}424 \ lb \qquad \text{(US/Brits)}$$

Liquid density of chlorine at its boiling point = 1,562 kg/m³

SI Units (Equation 7A)	*US/Brit Units (Equation 7B)*
$$A_p = 100\frac{W_p}{\rho_l}$$	$$A_p = 30.5\frac{W_p}{\rho_l}$$
$$A_p = 100\frac{19202}{1562}$$	$$A_p = 30.5\frac{42424}{97.5}$$
$$A_p = 1{,}229 \ m^2$$	$$A_p = 13{,}271 \ ft^2$$

Calculate the vapor flow rate from the pool.

Since chlorine is boiling in the pool, $P_v = 101.3$ kPa $= 14.70$ psi
Molecular weight of chlorine $= 70.91$

SI Units (Equation 8A)

$$AQ_p = 9.0 \times 10^{-4} \left(A_p^{0.95}\right) \frac{(MW)P_v}{T+273}$$

$$AQ_p = 9.0 \times 10^{-4} \left(1229^{0.95}\right) \frac{70.91(101.3)}{(-34.0)+273}$$

$$AQ_p = 23.3 \text{ kg / sec}$$

US/Brit Units (Equation 8B)

$$AQ_p = 0.154 \left(A_p^{0.95}\right) \frac{(MW)P_v}{T+459}$$

$$AQ_p = 0.154 \left(13271^{0.95}\right) \frac{70.91(14.70)}{(-29.2)+459}$$

$$AQ_p = 3,083 \text{ lb / min}$$

Calculate source strength of release.

SI Units (Equation 9A)

$$AQ = AQ_f + AQ_p$$

$$AQ = 38.8 + 23.3$$

$$AQ = 62.1 \text{ kg / sec}$$

US/Brit Units (Equation 9B)

$$AQ = AQ_f + AQ_p$$

$$AQ = 5139 + 3083$$

$$AQ = 8,222 \text{ lb / min}$$

Compare to the liquid release: 62.1 kg/sec is greater than 60.1 kg/sec and 8,222 lb/min is greater than 7,967 lb/min:

$$AQ = 60.1 \text{ kg / sec}$$

$$AQ = 7,967 \text{ lb / min}$$

Calculate the CEI.

SI Units (Equation 10A)

where ERPG-2 $= 9$ mg/m^3

$$CEI = 655.1 \sqrt{\frac{AQ}{ERPG\text{-}2}}$$

$$CEI = 655.1 \sqrt{\frac{60.1}{9}}$$

$$CEI = 1,963$$

This is greater than 1000; thus
$$CEI = 1,000$$

US/Brit Units (Equation 10B)

where ERPG-2 $= 3$ PPM

$$CEI = 281.8 \sqrt{\frac{AQ}{(ERPG\text{-}2)(MW)}}$$

$$CEI = 281.8 \sqrt{\frac{7967}{3(70.91)}}$$

$$CEI = 1,725$$

This is greater than 1000; thus
$$CEI = 1,000$$

Calculate the Hazard Distances.

SI Units (Equation 11A)

For ERPG-2 = 9 mg/m^3

$$HD = 6551\sqrt{\frac{AQ}{ERPG}}$$

$$HD = 6551\sqrt{\frac{60.1}{9}}$$

HD = 16,929 meters

HD is greater than 10,000 meters, thus

HD = 10,000 meters

For ERPG-1 = 3 mg/m^3

$$HD = 6551\sqrt{\frac{60.1}{3}}$$

HD = 29,321 meters

HD is greater than 10,000 meters, thus

HD = 10,000 meters

For ERPG-3 = 58 mg/m^3

$$HD = 6551\sqrt{\frac{60.1}{58}}$$

HD = 6,668 meters

US/Brit Units (Equation 11B)

For ERPG-2 = 3 PPM

$$HD = 9243\sqrt{\frac{AQ}{ERPG(MW)}}$$

$$HD = 9243\sqrt{\frac{7967}{3(70.91)}}$$

HD = 56,525 feet

HD is greater than 32,800 feet, thus

HD = 32,800 feet

For ERPG-1 = 1 PPM

$$HD = 9243\sqrt{\frac{7967}{1(70.91)}}$$

HD = 97,973 feet

HD is greater than 32,800 feet, thus

HD = 32,800 feet

For ERPG-3 = 20 PPM

$$HD = 9243\sqrt{\frac{7967}{20(70.91)}}$$

HD = 21,907 feet

CHEMICAL EXPOSURE INDEX FOR SELECTED CHEMICALS FOR RELEASES FROM A 2-INCH DIAMETER HOLE

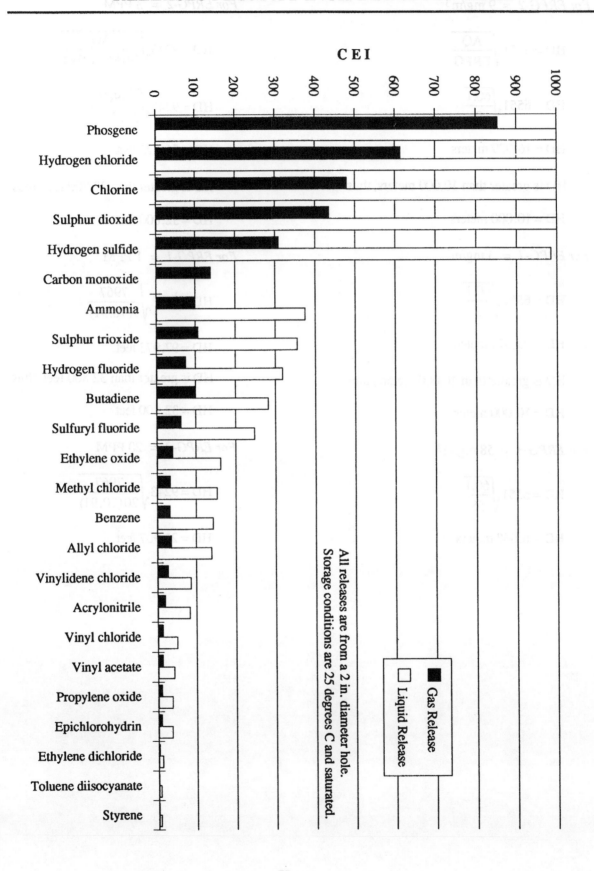

Printed and bound by CPI Group (UK) Ltd, Croydon, CR0 4YY

23/04/2025

14660912-0002